YOUR KNOWLEDGE HAS VALUE

David Brückner

Nuclear Fusion - Bringing the Sun to Earth

GRIN Verlag

Bibliografische Information der Deutschen Nationalbibliothek:

Die Deutsche Bibliothek verzeichnet diese Publikation in der Deutschen National-
bibliografie; detaillierte bibliografische Daten sind im Internet über http://dnb.d-
nb.de/ abrufbar.

Imprint:

Copyright © 2011 GRIN Verlag GmbH
Druck und Bindung: Books on Demand GmbH, Norderstedt Germany
ISBN: 978-3-656-35397-3

This book at GRIN:

http://www.grin.com/en/e-book/207643/nuclear-fusion-bringing-the-sun-to-earth

GRIN - Your knowledge has value

Der GRIN Verlag publiziert seit 1998 wissenschaftliche Arbeiten von Studenten, Hochschullehrern und anderen Akademikern als eBook und gedrucktes Buch. Die Verlagswebsite www.grin.com ist die ideale Plattform zur Veröffentlichung von Hausarbeiten, Abschlussarbeiten, wissenschaftlichen Aufsätzen, Dissertationen und Fachbüchern.

Visit us on the internet:

http://www.grin.com/

http://www.facebook.com/grincom

http://www.twitter.com/grin_com

NUCLEAR FUSION
BRINGING THE SUN TO EARTH

David Brückner

BCC Science Commendation 2011

Lancing College

TABLE OF CONTENT

1. Introduction

The dream that a bathtub of water and 100 g lithium could supply a family for 50 years with electricity[1] stimulated scientists since the 1940s all over the world to make every effort to construct a working fusion reactor that uses the most fundamental of all energy sources: the nuclear fusion that fuels sun.

However, the history of fusion research started a century earlier when scientists were trying to understand the process that fuels the sun.

2. Nuclear fusion in sun

A long time, it was an unsolved riddle to scientists, how the sun produces its energy. The sun can't simply be made of stone coal and oxygen, otherwise it have existed for about 5000 years.[2] Hermann von Helmholtz (1821-1894) proposed that it could heat up by consistent shrinking because gravitational energy can be converted into heat by contraction. If however the sun would have emerged from an interstellar nebula and would have contracted to its contemporary size, it would be roughly 22 million years old and geologists have proved that the earth is more than four billion years old and the sun can't be younger than its planet.[3] In 1920, Sir Arthur Stanley Eddington wondered wether the energy of stars was produced by fusion of hydrogen atoms to heavier atoms. And this principle of nuclear fusion is what we nowadays assume to be the motor of the stars including our sun.

In the core of the sun, the temperature is about 15 million °C and there is a pressure of 134 g cm^{-3}. The temperature quickly decreases to the outside: the radiative zone has a temperature of 3 million °C, the photosphere of 5500 °C. The average density of the sun is 1.4 g cm^{-3}, so most of the mass is concentrated in the core. Only in the core the conditions are such that nuclear fusion can take place as very high temperature and pressure is needed to let particles collide very often which allows their fusion.

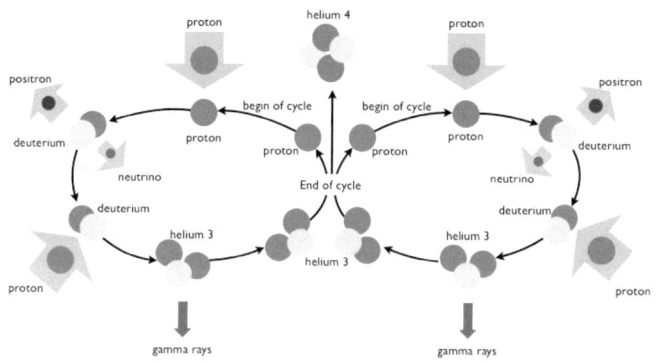

Fig. 1: The proton-proton cycle.

[1] *Die Fusion dringt zum Kern vor*, MaxPlanckForschung, No. SP, 2010, pp. 6-11, German

[2] Keller, Hans-Ulrich: *Das heiße Herz der Sonne*, in: Kosmos Himmels Jahr 2010, pp. 153-161

[3] John Wesson: *The science of JET*, 2000

Per cycle, 26.7 MeV are released. The conversion of a single gram hydrogen to helium gives an energy of 280,000 kWh. To maintain its luminosity, four million tons of matter are used for nuclear fusion each second. One more million ton is lost by the solar wind, a flow of charged particles such as protons, α-particles (a ^{4}He nucleus) and electrons that emanates from the sun. Despite these large quantities, the sun has not even lost 1% of its mass since its origin five billion years ago.

The release of energy by nuclear fusion can be explain using the relativity equation $E=\Delta mc^2$, where Δm is the mass loss and c the speed of light. A proton has a mass of 1.007825 u, a ^{4}He nucleus has a mass of 4.002603 u and the mass difference between four hydrogen nuclei and an α-particle is 0.028697 u. Using $E=\Delta mc^2$, the energy released is 26.57 MeV. The energy released is given in MeV, mega electron volts. 1 eV is equal to the amount of kinetic energy given to a single unbound electron when it accelerates through 1 volt of electric potential difference.[4] The e is just the elementary charge, therefore 1 eV $= 1.6 \times 10^{-19}$ J and the units cancel as follows: eV = CV = AsV = Ws = J.

3. Nuclear fusion on earth

In the late 1940's scientists began to investigate if it was possible to use the nuclear fusion, that had been discovered to be the sun's fuel, as an energy source on earth.

3.1 Fusion reactions

The source of fusion energy is the binding energy of the atoms. Splitting the nucleus of an atom into smaller parts requires energy; if two or more nuclei join together, energy is released. The most stable nuclei are those of iron, cobalt, nickel and copper which have an atomic mass of around 60. Nuclear fission of very heavy nuclei such as uranium, atomic mass 235, and fusion of light nuclei such as hydrogen and its isotopes ^{2}H (deuterium) and ^{3}H (tritium) can be used as energy sources. Nuclear fission is used since the early 1950s.

Fusion reactions

D + T	$\rightarrow$	^{4}He + n	17.58 MeV
D + D	$\rightarrow$	T + p	4.03 MeV
D + D	$\rightarrow$	^{3}He + n	3.27 MeV
D + ^{3}He	$\rightarrow$	^{4}He + p	18.35 MeV

D stands for deuterium, T for tritium, n for a neutron and p for a proton. The fusion of deuterium and tritium is the one with the highest reaction rate at the lowest temperature and has therefore the largest energy spoil.

---

[4] *Dictionary of Physics*, Oxford University Press, 2005

In this reaction, one deuterium nucleus reacts with one tritium nucleus to give a helium-4 nucleus and a neutron. This neutron carries 80% of the energy released. Deuterium can be obtained from sea water which for the required quantities is exhaustless and tritium can be obtained from the also very common element lithium by the following fusion reactions.

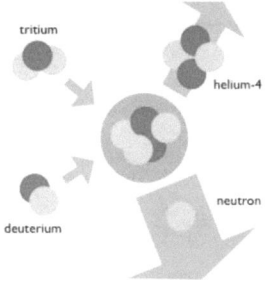

Fusion reactions in lithium

| $^7Li + n$ | → | $^4He + T$ | - 2.47 MeV |
| $^6Li + n$ | → | $^4He + T$ | 4.03 MeV |

Fig. 3: Schematic representation of the deuterium-tritium reaction.

3.2 The conditions of ignition

Only at very high pressure it is possible that the kinetic energy of the protons is high enough to overcome the electrostatic repulsion of equal charges so that they come so close to stick together. This is a distance of 10^{-15} m. It is impossible to create such a high pressure under terrestrial conditions but a temperature of 100 million °C has the same effect as it gives the atoms more kinetic energy.[5] The atoms are then in a state called plasma where electrons are separated from the nuclei which then form positive ions. In the case of hydrogen which has only one electron, it becomes fully ionised at a comparatively low temperature.

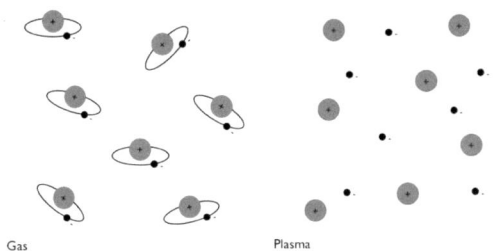

Gas

Plasma

Fig. 4, 5: Schematic representation of a gas and a plasma.

For ignition, three properties of the plasma are essential: the temperature, density and confinement time. Confinement time is a measure of the quality of the heat isolation of the plasma. The product of these three values needs to have a certain minimum value to allow ignition.[6]

---

[5] Wesson, John: *The science of JET*, 2000, http://www.jet.efda.org/wp-content/uploads/the-science-of-jet-2000.pdf

[6] Max Planck Institute for Plasma Physics: *Kernfusion - Berichte aus der Forschung*, 2003, German

4. The magnetic confinement

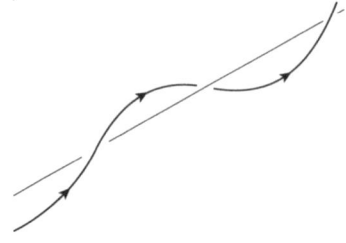

Fig. 6: The helical trajectory of a charged particle along a magnetic field

In the late 1940s it seemed an irresolvable problem to the scientist how to enclose the plasma as any contact to the reaction container wall would let the surface layer of the wall evaporate and cool the plasma rapidly and the fusion would shut down.[7] In 1951, Lyman Spitzer had the idea to enclose the plasma in a magnetic cage.[8] As the plasma is ionised, it consists of charged particles (positive ions and electrons) that can be influenced by a magnetic field. Their trajectory has two components: a circular motion at right angles to the magnetic field and a linear motion along the magnetic field.[9] Combined, the two parts of the motion give the spiral orbit shown in figure 8.

The scientists were enthusiastic and predicted in 1955 that in 20 years time, nuclear fusion would provide us with exhaustless energy.[8] However, the magnetic confinement turned out to be much trickier than assumed. The physicists were struggling to prevent particles to escape from the magnetic field.

In the 1950s, scientists all over the world tested a variety of confinement concepts and dismissed most of them. The two technologies that proved successful were tokamaks and stellarators.

4.1 The Tokamak

The tokamak was invented by the Soviet physicists Igor Tamm and Andrei Sakharov in 1952. Tokamak is an acronym of тороидальная камера с магнитными катушками (toroidal chamber with magnetic coils).[10]

Toroidal magnetic fields are used to avoid that the particles escape at the poles of the magnetic field. However, a toroidal magnetic field is not able to hold the plasma in an equilibrium force balance because the field strength decreases from the inside to the outside of the toroidal field with the effect that the particles drift towards the wall. Therefore, the field lines may not take a circular course about the the axis of the torus but need to be helically looped.[11]

[7] EFDA JET: *Cleaner energy for the future - The development of fusion power*, 2006, p. 2

[8] *Vielleicht-Maschine mit Zukunft*, MaxPlanckForschung, No.2, 2010, pp. 90-91

[9] John Wesson: *The science of JET*, 2000, p. 10

[10] John Wesson: *The science of JET*, 2000, p. 13

[11] Max Planck Institute for Plasma Physics: *Kernfusion - Berichte aus der Forschung*, 2003, pp. 10-11

To create the helical field lines, two magnetic fields are superimposed: The toroidal field produced by the field coils and the poloidal field of an additional electric current that flows through the plasma.

The current is produced by the transformer coil in the centre of the torus (see figure 7). The transformer cannot produce the increasing current continuously so a tokamak can only work in pulses. In a future fusion power plant, they will probably last about 1 hour. However, power plants in the mains may not work pulsed for technical reasons, methods are explored how to create a continuous current.

4.2 The Stellarator

The principle of the stellarator can be traced back to Lyman Spitzer. The stellarator is much more complex than the tokamak but it needs no transformer current which allows a continuous running of the reaction. This is because no superposition of a poloidal and toroidal field is required to create a helical magnetic field because the magnetic coils are twisted in a way that the particles cannot escape.

The symmetry of stellarators are very complicated, different helical and axial symmetries are possible.

5. Nuclear fusion as energy source

All this research on nuclear fusion and plasma physics has been done because scientist are seeking to use this as an energy resource. The long and costly research has not been given up so far because the possible results are so tempting: For the reaction, only deuterium and tritium are required. Deuterium can be obtained from sea water and tritium is being produced in the reactor itself from lithium which is a common element in minerals.[12]

5.1 The fusion power plant

Lithium and deuterium needs to be brought in. The lithium will react to give tritium (see section 2.1). The only waste product is helium. The energy is transmitted by the high-energy neutrons that released in the fusion reaction. They are neutral so will not be influenced by the magnetic cage. So they leave the plasma and hit the blanket. Their kinetic energy is transformed into heat which is lead off to produce electric current in the turbine generator.

5.2 Application of fusion power in the energy system

A fusion plant would give an electric power of about 1500 megawatts and it would serve the base load like contemporary large-scale power plants. In the near future, it will be necessary to switch to renewable energies such as solar and wind power. However, these energy sources are mostly weather dependent so it is preferable to have continuously running power plants in addition. The contemporarily used technologies are far less environmentally friendly than fusion power.

5.2 Safety and environmental issues

A major advantage of nuclear fusion to contemporary large-scale power plants is that it does not use up resources (the required quantities of water and lithium are exhaustless) and that it does not produce greenhouse gases like coal-fired power plants. The helium produced is not harmful, it occurs

---

[12] *Vielleicht-Maschine mit Zukunft*, MaxPlanckForschung, No. 2, 2010, pp. 90-91, German

naturally in the atmosphere. Compared to nuclear fission, its radioactive waste is very short-lived and an accident would not cause a large-scale catastrophe.

The main safety concern about nuclear fusion is the radioactive tritium and the blanket that is activated by high-energy neutrons.

In the power plant about 1 kg tritium will be present. Hitherto existing experiences it is assumed that about 1 g tritium will escape per year.[13] However, the half-life of tritium is only 12.32 years which is very short compared to elements used in fission plants (e.g. ^{235}U has a half-life of 7×10^8 years). The concentration of tritium in soil decreases even faster. The following graph[14] shows the drop off of tritium in soil. After about 100 days the radiation is not much more that the background radiation (0.1 Bq kg^{-1}).

The half-life of the the activated materials in the blanket is in the range of a few months to several years, not much compared to ^{235}U and much less problematic to store.

Accidents like in Tchernobyl and Fukushima are not possible as no chain reactions such as that of ^{235}U cannot occur.

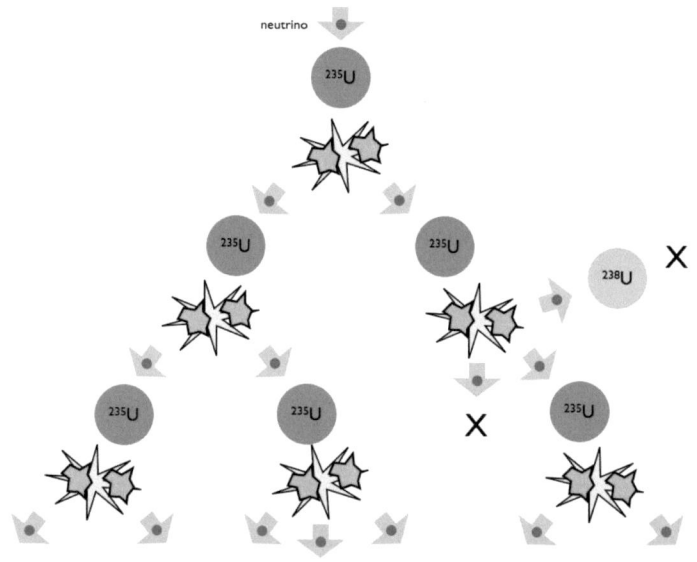

Fig. 16: Chain reaction of ^{235}U in fission power plants.

These chain reactions are very dangerous if they get out of control. In a fusion reactor, no chain

---

[13] Max Planck Institute for Plasma Physics: Kernfusion - Berichte aus der Forschung, 2003, p. 32

[14] Source of data: Brown et al.: Field Studies of HT Oxidation and Dispersion in the environment II - The 1987 June Experiment at Chalk River, Canadian Fusion Fuels Technology Project, CFFTP-G-88007, Ontario Hydro, Canada, October 1988

reactions can occur. Moreover, if the plasma vessel would leak, the plasma would shut down immediately because even small amounts of entering air would cause the plasma fire to extinguish.[15]

6. The progress of fusion research

Fusion research has come close to ignition now. The european joint project JET in Culham, UK has contributed significantly to this success. Most of the plotted reactors are tokamaks, only the German Wendelstein series and the Japanese LHD have explored the potential of stellarators.

At the moment the international joint project ITER is being built in Cadarache, Souther France. It will hopefully allow the first ignition to happen. If this proves successful, demonstration power plant will be constructed, DEMO. According to the plans of fusion researchers, a working power plant will be existent in around 2050.[16]

---

[15] Max Planck Institute for Plasma Physics: *Kernfusion - Berichte aus der Forschung,* 2003, p. 19

[16] *Research for the energy of the future,* 2011, http://www.ipp.mpg.de/ippcms/de/pr/publikationen/pdf/broschuere_engl.pdf

REFERENCES

Dictionary of Physics, Oxford University Press, 2005

EFDA JET: *Energy - Powering your world*, 2002, http://www.jet.efda.org/wp-content/uploads/Energy-Powering-Your-World-2002.pdf

EFDA JET: *Cleaner energy for the future - The development of fusion power*, 2006, http://www.jet.efda.org/wp-content/uploads/EFDA-Cleaner-Energy-for-the-Future-2006.pdf

Fullick, Patrick: *Physics*, Heinemann Advanced Science, 2003

Keller, Hans-Ulrich: *Das heiße Herz der Sonne*, in: Kosmos Himmels Jahr 2010, pp. 153-161

Lackner, Karl: *1958 - 2008: 50 Jahre Fusionsforschung für den Frieden*, in 50 Jahre Plasmaphysik und Fusionsforschung in Innsbruck, 2008

Lexikon der Physik, Deutscher Taschenbuch Verlag, 1970, German

Maisonnier, David et al.: *A conceptual study of commercial fusion power plants*. Final Report of the European Fusion Power Plant Conceptual Study (PPCS), 2005, http://www.efda.org/eu_fusion_programme/downloads/scientific_and_technical_publications/PPCS_overall_report_final.pdf

Milch, Isabella: *Die Fusionsanlage ASDEX Upgrade – ein europäisches Forschungsinstrument*, in Brains and Tools 2004, pp. 21-24

Milch, Isabella: *Sonnenfeuer im Labor - Wo steht die Fusionsforschung?*, Kultur & Technik, Deutsches Museum München, No. 2, 2007, pp. 44 - 49

Max-Planck Institute for Plasma Physics, Munich, Germany:

> *Kernfusion - Berichte aus der Forschung*, 2003, http://www.ipp.mpg.de/ippcms/de/pr/publikationen/pdf/berichte.pdf

> *Research for the energy of the future*, 2011, http://www.ipp.mpg.de/ippcms/de/pr/publikationen/pdf/broschuere_engl.pdf

> *Max-Planck-Institut für Plasmaphysik. 50 Jahre Forschung für die Energie der Zukunft*, 2010, http://www.ipp.mpg.de/ippcms/de/pr/publikationen/pdf/50_Jahre_IPP.pdf

> *Wendelstein 7-X fusion experiment*, 2008, http://www.ipp.mpg.de/ippcms/eng/pr/publikationen/W7X_engl.pdf

> *ASDEX Upgrade fusion experiment*, 2006, http://www.ipp.mpg.de/ippcms/eng/pr/publikationen/AUG_engl.pdf

> *Nuclear fusion - Status and Prospects*, 2011, http://www.ipp.mpg.de/ippcms/eng/pr/publikationen/fusion_e.pdf

> Hasinger, Günther: *Stand der Fusionstechnik*, 2010, http://www.ipp.mpg.de/ippcms/de/pr/publikationen/pdf/Hasinger_KWTKOLL.pdf, German

> *Magnetfelder bändigen die Urgewalt*, MaxPlanckForschung, No. 2, 2006, pp. 26-30, German

> *Vielleicht-Maschine mit Zukunft*, MaxPlanckForschung, No. 2, 2010, pp. 90-91, German

> *Die Fusion dringt zum Kern vor*, MaxPlanckForschung, No. SP, 2010, pp. 6-11, German

Wesson, John: *The science of JET*, 2000, http://www.jet.efda.org/wp-content/uploads/the-science-of-jet-2000.pdf